Appendix to the American Telegraph Magazine.

FIRST TELEGRAPH CASE

BEFORE THE

UNITED STATES SUPREME COURT.

HENRY O'RIELLY AND OTHERS,
APPELLANTS,
versus
S. F. B. MORSE AND OTHERS,
APPELLEES.

SKETCH OF THE

OPENING ARGUMENT OF R. H. GILLET,

ON THE

APPEAL OF O'RIELLY FROM THE DECISION OF JUDGE MONROE IN KENTUCKY.

DECEMBER 24TH AND 27TH, 1852.

NEW-YORK:
PRINTED BY JOHN A. GRAY,
97 CLIFF, COR. FRANKFORT ST.
1853.

Appendix to the American Telegraph Magazine.

FIRST TELEGRAPH CASE

BEFORE THE

UNITED STATES SUPREME COURT.

HENRY O'RIELLY AND OTHERS,
APPELLANTS,
versus
S. F. B. MORSE AND OTHERS,
APPELLEES.

SKETCH OF THE

OPENING ARGUMENT OF R. H. GILLET,

ON THE

APPEAL OF O'RIELLY FROM THE DECISION OF JUDGE MONROE IN KENTUCKY.

DECEMBER 24TH AND 27TH, 1852.

NEW-YORK:
PRINTED BY JOHN A. GRAY,
97 CLIFF, COR. FRANKFORT ST.
1853.

F. C. Wagner

ARGUMENT

ON THE

APPEAL OF O'RIELLY TO THE U. S. SUPREME COURT, &c.

THIS cause comes here from the Kentucky Circuit, and involves two questions: the validity of Professor Morse's patents, and the identity of his inventions with Barnes' and Zook's Columbian Telegraph, used by O'Rielly. I shall not attempt a history of the rise, progress, and perfection of telegraphic inventions, because it would not, more than the exhibition and lecture proposed on the other side, be found in the record, or aid in determining the legal questions in the case. I shall not controvert Morse's merit as an inventor, while I claim that another useful, and in some respects superior, improvement has been made, which O'Rielly used. But justice requires me to state, that without the great development of electro-magnetic power, discovered by Professor Henry of this city, Morse's invention of applying that power would never have been made. I must also add, but for the skill and perseverance of a most valuable class of men, the telegraphic operators, I fully believe Morse's telegraph would never have been so perfected, and made practically useful. If they have not patents to commemorate the products of their genius and industry, or to enrich themselves, they are none the less meritorious. Their services to mankind are not lessened because their inventions are now embodied in Morse's reïssued patents, swelling his reputation, and increasing his wealth.

I shall not compare Dr. Jackson's and Morse's claims to the invention in 1832, because nothing was then so perfected as to be patentable. Nor is there evidence in this case to show that Morse completed his invention prior to April 8th, 1838. On the 6th of the previous October he filed a caveat, to entitle him to notice if another should apply within a year, upon the alleged ground, "that the machinery for a full practical display of his new invention is not yet completed, and he therefore prays protection of his right until he shall have matured the machinery." Surely, when he had not matured machinery to display his invention, it was not in a condition to be patentable. Hence I shall disregard the speculations of 1832.

This suit is brought on Morse's patent of 1840, reïssued in 1846, and again in 1848, and on his patent of 1846, reïssued in 1848. The bill is clearly multifarious, for including two patents, thereby complicating the trial; but we did not raise that objection in our pleadings, and cannot now do so here, though the Court has the power to interpose it, on its own account. Morse charges O'Rielly with having constructed and used a telegraph in Kentucky, in violation of his (Morse's) patents, by recording signs and making sounds. O'Rielly denies the infringement, and alleges that his (Morse's) patents are invalid for various causes, enumerated in his answer.

A preliminary injunction was granted. Subsequently messages were sent and recorded in Tennessee, and the answers were gathered in Kentucky, by listening to and interpreting the click made by armature when striking against the magnet or stops. For this acoustic offence the operator was arraigned for a contempt of court. Afterwards O'Rielly's transmitting station was removed from Louisville to Jeffersonville, in Indiana, and messages were sent and received over the wires, which extended across Kentucky. These harmless pulsations of the galvanic current, passing between Indiana and Tennessee, were also deemed a contempt. To prevent future violations of law, and further disrespect of judicial proceedings, the Court, by its mandate, directed the Marshal of Kentucky to take possession of the line, and to cut the wires. The Marshal obeyed, and the guillotine was applied to the sinning wires. After the execution, the galvanic lightning ceased its sinful contempts, and judicial dignity was permitted to repose in the enjoyment of all the respect to which its action was entitled. This feat, probably, has not its parallel in the history of judicial proceedings. Decapitating the wicked to prevent their committing sin would not go far beyond it. While I say this, I am bound to say, that the Judge whose acts I am criticising is a learned, pure, and honest man. Nine months after the execution of the wires, on giving good bail, the lightning current was so far forgiven its offences, as to be allowed to flash across Kentucky without fear of the Marshal.

The final decree declared Morse's patents valid, and a perpetual injunction during their lives was ordered. O'Rielly's telegraph was declared to be an infringement, and he was forbidden to use it or one like it, "in the transmission of intelligence, which is in one place, to another distant place, by making thereat a permanent record," or "by making thereat, with the action of the instrument, which would make such characters, *alphabetical sounds*, and out of them, composing such characters." Although alphabetical sounds are not named in Morse's patents, nor a mode of making or using any sounds, except to sound an alarm bell, as Wheatstone had previously done, still, as the operator had heard sounds, and had contrived to interpret them, Morse's monopoly would not be perfect, unless he monopolized sounds, as well as electro-magnetic power. Hence the Court, being misled, was induced to restrain O'Rielly from doing what was not, in any manner whatever, covered by Morse's original patents, or in either of their frequent and mysterious expansions. There being no apology in the patents for this part of the decree, it cannot stand. I leave this point, and go to the other great questions.

The telegraph now used by Morse, and included in his patents, consists of the following things:—

1. *A galvanic battery*, consisting of a zinc and copper plate united by a wire, placed in a cup of nitric acid. That principally used is Groves' battery, invented in 1839. Others had been long known.

2. *Metallic conductors* of copper or iron wire, connected, the one to the zinc, and the other to the copper plate of the battery, insulated on posts in the air. The use of the wires is an old invention. Steinheil used posts in 1837. In 1843 Morse buried his wire in pipes in the ground, but on advice by Professor Henry and others, resorted to the posts.

3. *The ground circuit.* The opposite poles of a battery, here and at Baltimore, are connected by wires to zinc plates in the ground at both places, and with the wires on the poles in the air, and thus a circuit is formed, the earth making a part. This use of the earth was discovered and used by Steinheil in Bavaria in 1837. Morse commenced by using two wires to make the circuit, and subsequently used the earth. The electric current will not flow from one pole of a battery, until means are provided for its return to the other.

4. *A finger key.* This was used by Ampère in 1820. Morse at first used saw-teeth type, a port rule, a forked wire and cup of mercury to plunge it

in, to accomplish the object attained by Ampère in the use of the finger key, which is more simple and convenient.

5. *The electro-magnet,* which attracts a bar of iron. Sturgeon, in England, contrived, in 1825, what was called an electro-magnet. In 1829 Professor Henry invented the powerful and practically useful electro-magnet, now used. It consists of numerous coils of small isolated wire, wound around a bar of soft iron in the form of a horse-shoe. When the galvanic current flows from a battery, it collects in these coils, and the iron bar becomes highly magnetic, and attracts soft iron within its influence. Without this embodier of the current, a sufficient quantity of galvanism could not be transmitted, to become effective by attraction. It is the attractive properties of this electro-magnet when near iron, which constitute the electro-magnetic power used by telegraphers. Morse did not first discover this power. Henry perfected the mode of using it, and rendered it available, for which the world owes him a debt not easily cancelled. His only reward is the exalted reputation it confers.

This power and the means of creating it were public property before Morse's invention. He applied the former to telegraphing by adding,

6. *A recording apparatus.* This consists in so shaping the bar of soft iron, (called an armature,) which is attracted by the electro-magnet, that when it moves toward the electro-magnet, one end, having a point in it, is made to come in contact with a fillet of paper moved by clock-work. When the galvanic circuit is closed by pressing down the finger key, the current collects in the electro-magnet, and attracts the bar of soft iron, or armature, which raises the end containing the point with such force as to indent the paper. When the circuit is broken, the galvanic current vanishes from the electro-magnet, and the latter no longer attracts the iron bar, or armature, and the paper moves on without being indented. The frequency and length of these indentations depend upon the length of time the circuit is opened and closed. These indentations are variously combined, and thus form an alphabet, which operators soon learn. Morse's invention consists solely in this recording apparatus, and arranging it in connection with the circuits. More than that he did not invent. Every thing up to that point was perfect and public property, to be used by those who chose, before his invention. The opening and closing one circuit by another, though patented by Morse, was indisputably invented by Professor Henry years before, and used by Wheatstone before Morse filed his caveat. Not content with the great merit to which this invention entitled him, the American press has teemed with articles, prepared by him or his friends, and thereby the public has been erroneously led to believe that he invented the whole telegraphic apparatus. A false public opinion has been created, which time and judicial inquiry will hereafter correct. History will record, that the telegraph has been produced by the agency of numerous individuals, and that Morse's share in it is not greater than Steinheil's or Henry's. If equal to either, it should satisfy the ambition of even an ambitious man. But Morse's claims are almost unbounded, and centre in the desire of a rich harvest from his patents. Although he did not first discover the motive power of electro-magnetism, he claims its exclusive use, when applied *to make a telegraphic record, or to produce sounds;* that is, he claims all possible modes, when he only invented one mode of applying it. If the law is as he claims it to be, if I invent a mode unknown to him, and infinitely better than his, and use the power of electro-magnetism, which he did not discover, he can use my invention at his pleasure, and can sue and recover against me if I use my own invention. This would be a palpable infringement of the Constitution, which authorizes laws granting the exclusive use by an inventor of his own, and not of another man's, invention. To allow him to recover against me for using my own invention, would be an outrage upon the law, upon common sense, and upon

justice itself. It will require more than the power and authority of our own pure and venerated judiciary to secure the acquiescence of the American people in such an extraordinary conclusion. A claim to the exclusive use of air, fire, water, and steam, to produce a particular result, would have the same foundation.

Morse's patent, as first issued in 1840, only claimed a certain mode of applying the power of galvanism, or electro-magnetism, and did not claim its exclusive use. In the second reïssue of this patent, a few months after his agents and friends had seen permanent magnetism used in the O'Rielly instrument, he was allowed to claim the exclusive use of "*the motive power of magnetism.*" Morse's instrument can only be worked by electro-magnetism. The Columbian is operated by the employment of two permanent magnets as the motive power. Electro-magnetism is not used in it as a motive power at all, but simply to change the polarity of an iron bar, so as to release or subject it to the power of the permanent magnets. These are not as much affected by atmospheric electricity as the electro-magnet. The testimony shows the instruments to be substantially different; and not a witness swears that the Columbian is an infringement, unless he did so under the belief, that to use magnetism as a motive power, was a violation of Morse's patent. This is either expressly stated, or inferable from the testimony of Morse's witnesses. Those called by O'Rielly clearly show that the two inventions are not alike. We confidently assert that there has been no infringement, because the inventions are unlike in form and structure, and in the power used, and in the mode of its application, and that O'Rielly has used nothing that Morse in fact invented. Before this Court can hold that O'Rielly infringed, it must determine that Morse is entitled to the exclusive use of permanent, as well as electro-magnetism, although differently produced and applied. He has shown no mode of using it, and it cannot be used by any machinery or combination invented by him. It will require more inventive genius than Morse has ever manifested, to produce substantial reasons for determining that these inventions are substantially the same.

I now proceed to other questions.

First. Morse's patent of 1840 is void, because it runs fourteen years from the date of its issue, instead of that length of time from the date of his patent granted in France in 1838, for the same invention.

The seventh section of the act of 1836 prohibits the issue of a patent where one has been previously granted for the same thing. The eighth section modifies this in favor of the true inventor, if, within six months after obtaining letters abroad, he files his specifications and drawings here. The sixth section of the act of 1839 suspends so much of the seventh section of the act of 1836 as relates to patents obtained abroad, with the proviso, "that in *all* cases, every such patent shall be limited to the term of fourteen years from the date or publication of such foreign letters patent." The object of this section was to limit patents to fourteen years from the time the inventor first patents the invention, either here or abroad. Under the act of 1836, a patent could run only fourteen years from the time of patenting here, which must have been within six months from the date of the foreign patent. The public would then have had the right to use the invention at the expiration of fourteen years and six months from the date of the foreign patent. Under the construction claimed by Morse, he would be entitled to fourteen years here, though he had had it fourteen abroad, and it had become public property there. This would place us fourteen years behind foreign nations in the right to use a patented invention. Morse claims that the act of 1839 does not reach his case, because he had applied for a patent before it was passed. We deny that his application was a valid one, within the meaning of the law, until he perfected it in 1840. From 1838 it was imperfect, and not conformable to law, and remained unacted upon at his written request, and was therefore no legal application. But if it was, the act makes no exceptions in favor of such a case, or any other one. His was as much within the mischief intended to be corrected, as any case; and the remedy must apply. For this cause, we say this patent is invalid.

Second. In construing patents, and in deciding what the inventions are that are embraced in the patent, the summing up is conclusive, and nothing is patented which is not expressly claimed.

This doctrine is settled in England. In Rex *vs.* Cutler, 1 Starkie R. 354; Bovil *vs.* Moore, 2 Marsh R. 211, this rule was laid down. The elementary authors are to the same effect. Judge Story, in Moody *vs.* Fiske, 2 Mason 112–118, lays down the same rule. In Wyeth *vs.* Stone, 1 Story R. p. 285, he says: "In order to ascertain the true construction of the specification in this respect, we must look to the summing up of the invention, and the claim therefor asserted in the specification; for it is the duty of the patentee to sum up his invention in clear and determinate terms. This was the doctrine maintained in Moody *vs.* Fiske, 2 Mason 112, and I see no reason to doubt or depart from it." In Hovey *vs.* Bemis, 3 Woodbury and Minot 17, this doctrine is fully confirmed. The sixth section of the act of 1836 provides: "And he shall particularly specify and point out the part, improvement, or combination which he claims as his own invention or discovery. This being a statutory requirement, Morse is bound by his acts under it. Settling this position will enable me hereafter to show what Morse has legally included in his several patents.

Third. Whatever is described in Morse's patents and not claimed in the summing up, whether invented by him or not, is abandoned or dedicated to the public, and cannot be afterwards claimed, as a part of his patent in the reïssue, or otherwise.

This proposition is in part the same as the last, but it also declares the effect, or consequence, of describing what is not claimed. The object of the statute in requiring the patentee to specify and point out what he claims, was to protect the patentee against the hazard of being misunderstood as to his invention; and also to enable the public to ascertain what could be lawfully used, and what was prohibited. The printed brief shows the numerous changes that Morse's patents have undergone; and the things described in the first issues and not claimed, but are afterwards included. The alphabet is a specimen. It is claimed in the last, but not in the first patent. Also the exclusive use of electro-magnetism and magnetism. It is as much described in 1840 as in the reïssue of 1848, but it is not claimed in the former. So of numerous things which can only be seen and understood by closely scrutinizing and comparing the original patents with the reïssues.

Morse insists upon the right to claim all that he described, but omitted in the summing up of the original patent. We say that he is concluded by his acts, and that his intentions in relation to these matters, about which we can know nothing, have no bearing upon the case. The public can observe his acts and draw their inferences, and ought not to be prejudiced by his blunders, intentions, or change of purpose. In Battin *vs.* Taggart, Judge Kane held "that a description by the applicant for a patent of a machine, or a part of a machine, in his specification, unaccompanied by notice that he has rights in it as an inventor, or that he desires to secure title to it as patentee, is a dedication of it to the public." "Neither of these sections authorizes a change in the character of the claim, the substitution of a different patentable subject. The defect to be remedied is, either some insufficiency in description, or an error in claiming too much."

In Grant *vs.* Raymond, 6 Peters 248, this court, quoting from Pennock *vs.* Dialogue, 2 Peters 1, says, "It is possible that the inventor may not have intended to give the benefit of his discovery to the public." "But it is not a question of intention," "but of legal inference, resulting from the conduct of the inventor, and affecting the interests of the public." This is reiterated in the opinion delivered in Shaw *vs.* Cooper, 7 Peters 318, where this court again quotes from Pennock *vs.* Dialogue, and says, "It has not been, and indeed cannot be denied, that an inventor may abandon his invention and surrender or dedicate it to the public. This inchoate right, thus gone, cannot afterwards be resumed at his pleasure; for when gifts are once made to the public in this way, they become absolute."

These cases settle the question in our favor. By not claiming the things referred to, Morse dedicated them to the public use, and when once gone, they cannot be recalled. Justice requires this should be so. Before including things in the claim in his patent, it was lawful for any one to use them, and millions might have been invested for that purpose. It would be wrong to allow him, by his own acts, to change these lawful investments into unlawful ones. Public policy, as well as private right, requires that he should be bound by his acts. If this doctrine is applied to Morse's patents, they must fall, for including in the reïssues what belonged to the public when claimed.

Fourth. A patent void in part is void in whole, except when otherwise provided by statute.

A patent is an entirety, and if void in one part it must be wholly so. This was so at common law, and is so by statute. There are numerous defects which will defeat a patent. A disclaimer properly made and recorded will remedy a class of defects in certain cases. But here there has been no disclaimer. If the patent includes what was not lawfully patentable, it vitiates it. In Wyeth *vs.* Stone, 1 Story R. 285, in Moody *vs.* Fiske, 2 Mason 118, and in Woodcock *vs.* Parker, 1 Gall 437, Judge Story held if the claim was too broad the patent was void. In Bovil *vs.* Moore, Ch. J. Gibbs held the same. In Evans *vs.* Eaton, 7 Wheat. 356, this court said, "If therefore the patent be for the whole machine, the party can maintain title to it only by establishing that it is substantially new in its structure and mode of operation."

The suggestion that Morse may disclaim within a reasonable time, which had not elapsed when this suit was brought, is easily answered. We have, by statute, a right to take issue upon that question, but as he had not disclaimed, we have had no opportunity to do so. This court cannot know that Morse will ever disclaim. It is certain he will not, if he can succeed here without doing so. In Wyeth *vs.* Stone, Judge Story held, "If the court should grant a perpetual injunction before any disclaimer is filed, it may be that the patentee may never afterwards, within a reasonable time, file any disclaimer, although the act certainly contemplates the neglect or delay to do so to be a good defence upon a patent to secure the rights granted thereby." As Morse now has in his patents what he could not lawfully claim, they must be bad. He has claimed some things that were public property, and others, like the alphabet, which are not patentable, as well as the exclusive right to use the motive power of electro-magnetism to print letters, when he has shown no mode by which the printing can be done. For these causes, these patents, and especially that of 1840, must be declared invalid.

Fifth. An invention is not complete, so as to be patentable, or to bar the obtaining a patent by another inventor, until it is perfected and adapted to use.

The time when Morse's invention became patentable is highly important, with reference to the question whether he was the first inventor, as well as with regard to whether his invention had been patented or described before his application. Before we can fix the time of its being patentable, we must ascertain when the law deems things sufficiently perfected. As the patent must be for the machine or product, and not for the conception or idea, no patent can be granted until such machine or product is completed so as to be described in a specification, and represented in a model and drawings I have shown that Morse had not, when he filed his caveat in October, 1837, so far completed his machine as to enable him to do either. He then had no completed machine. And there is no evidence in this case of his making any further progress, until it is shown in his application in April, 1838, which is the true date of the invention. Prior thereto he had nothing that was in a condition to be patented. In Reed *vs.* Cutter, 1 Story 590, Judge Story lays down the true rule in these words: "He is the first inventor in the sense of the act, and entitled to a patent for his invention, who has first perfected and adapted the same to use; and until the invention is so perfected and adapted to use, it is not patentable. An imperfect and incomplete invention, resting in mere theory, or in intellectual notion, or in uncertain experiments, and not reduced to practice, and embodied in some distinct machinery, apparatus, manufacture, or composition of matter, is not, and indeed cannot be, patentable under the patent acts." The same doctrine is laid down in Washburn *vs.* Gould, 3 Story R. 133, and in Allen *vs.* Blunt, 2 Wood and Min. 121–141. Morse's invention not being in a condition to be patented before April, 1838, it is perfectly certain that he was preceded by Steinheil and by Wheatstone. The former brought his invention into practical use on a line still existing between Munich and Bogenhausen, in Bavaria, in 1837, where he recorded by ink or pencil, as Morse did afterwards. And Professor Henry distinctly swears that in April, 1837, he saw Wheatstone's apparatus, with the contrivance of bringing into operation a second circuit by means of the first, by the power of electro-magnetism. This places it beyond all doubt that Steinheil and Wheatstone had perfected their inventions before Morse did his. This being so, Morse was not, either in fact or in contemplation of law, the first inventor of what he claims. He was preceded in England and in Bavaria.

Sixth. Where a patent is for a combination of different parts, and not for the parts themselves which compose it, if the patentee did not invent the whole combination, but merely made additions to an old one, his patent is invalid. And where his patent is for a combination of parts, the use of those parts, if less than the whole, is no infringement.

Morse, in his patents themselves, admits that others had preceded him in the invention of telegraphs, whose motive power was electro-magnetism. But he claims that his was the first recording telegraph. From the galvanic battery to the electro-magnet, attracting a bar of iron, turning upon a pivot, or moving on its axis, including wires and the means of opening and closing the circuit, and the electro-magnet itself, which is the all-important and vital thing, Morse has admitted in the historic parts of one of his patents that it existed previously, and it has been proved in this case, that up to that point, the combination was old and in use before his invention. So much of the recording apparatus as consists of weights and clock-work is also proved to be old. Notwithstanding all this, he has claimed the whole of these combinations in his patent of 1840. His second claim is for the "employment of the machinery called the register or recording instrument," most of which is an old combination. In his third claim he says, "I also claim as my invention and improvement the combination of the machinery herein described, consisting of the generator of electricity, the circuit of conductors, the contrivance for breaking and closing the circuit, the electro-magnet, the pen or contrivance for marking, and the machinery for sustaining and moving the paper, all together constituting one apparatus, or telegraphic machine." This is one combined whole. By his own recitals in his patents, and by the undisputed testimony in this cause, this combination, from the generator of electricity to and including the electro-magnet and armature, existed, and was an old combination before Morse's invention. The pen, or the armature and machinery for moving the paper, were his only additions; yet he has claimed the whole, and a too confiding public has been led to believe he invented it. In his fourth claim he says, "I also claim as my invention or improvement, the combination of two or more galvanic or electric circuits, with independent batteries," &c. This combination is proved, by Professor Henry, to have been invented by him years ago, and he saw it in use by Wheatstone, in April, 1837. He has not the merit of making even an addition to it. Claiming a combination which he did not wholly invent, vitiates his patent. So, if he claimed a part which he did not invent. This court, in Evans *vs.* Eaton, 7 Wheaton 356, said, "If his patent includes things before known, or before in use, as his invention, he is not entitled to recover, for his patent is broader than his invention. If therefore the patent be for the whole machine, the party can maintain title to it only by establishing that it is substantially new, in its construction and mode of operation. If the same combination existed before in machines of the same nature, up to a certain point, and the party's invention consists in adding some new machinery, or some improved mode of operation of the old, the patent should be limited to such improvement; for, if it includes the whole machinery, it includes more than his invention, and therefore cannot be supported." Unless this court now overrules this solemn and well-considered adjudication, it must hold this patent invalid for being broader than the invention. The defendants below used no such combination as is included in either of Morse's claims, as the proof clearly shows; and if they did not, they are not liable, even if Morse's patents are good. This point has been repeatedly held by Justices Story and McLean on their circuits. This court, in Prouty *vs.* Ruggles, 16 Peters 336, declared, "The use of any two of these parts only, or of two combined with a third, which is substantially different in form, or manner of its arrangement and connection with the others, it is therefore not the thing patented. It is not the same combination if it substantially differs from it in any of its parts." That the Columbian differs from Morse's invention "in form or manner of its arrangement, and connection with other parts," is too clearly proved to be controverted. Hence Morse cannot recover for a violation of the combinations he has claimed. If he recovers at all, it must be because he is entitled to the exclusive use of magnetism as a motive power, the right to which I shall hereafter discuss.

Seventh. Morse's patents of 1846 and 1848 are void, because he was not the first inventor of the things patented, or of substantial and material parts thereof.

I have already shown that Morse's invention was subsequent to Steinheil's and Wheatstone's. The invention of the former included even the recording of signs, and the latter all that Morse uses, except he made intelligible signs by the electro-magnet attracting or deflecting a needle, instead of actually making a record upon paper. Morse answers these prior inventions, by alleging that he is the first American inventor, and as such, he is entitled to a patent, even if the invention had been previously made and used abroad. We say, that to sustain his patent, he must be the first inventor, as to all the world. In Pennock *vs.* Dialogue, 2 Peters 1, this Court said, "It (the 6th section of the act of 1793) gives the right to the first and true inventor, and to him only; and if known and used before his supposed discovery, he is not the first, although he may be a true inventor." This was in 1829, and in 1833, the case of Shaw *vs.* Cooper, 7 Peters 292, came up, and the same doctrine was reiterated. In this case there had been a use abroad, and this Court said, "If the invention was known or used by the public before it was patented, the patent is void." Judge Washington had held the same in four, and Judge Story in five cases. Judge Baldwin had held, "The invention must be new as to all the world." From 1804 to 1833, the decisions had been uniformly the same. The question now is, did the statute of 1836 change the law? We say it did not. The sixth section of the act of 1836 provides that a person who has made an invention which was "*known* or *used* by *others*, before his" invention, may apply for a patent on making oath, "that he does verily believe that he is the original and *first* inventor." This is the old law, and was actually followed in procuring the patents now under consideration. The seventh provides, that if on examination it "shall not appear to the Commissioner that the same has been invented or discovered by any other person in this country prior to the alleged discovery thereof by the applicant, if he shall deem it sufficiently useful and important, it shall be his duty to issue a patent therefor." The section then provides, if "on such examination it shall appear to the Commissioner that the applicant was not the original and *first* inventor thereof, or that any part of that which is claimed as new has been invented or discovered in this, or a foreign conntry, as aforesaid," then he is not to issue one, and certain proceedings are to be had before a board. The last clause in this section, being clear and express, must control the first, if there is a conflict between them. This last clause, and the provision in the sixth section, are perfectly harmonious. The fifteenth section provides, if "the patentee was not the original and *first* inventor or discoverer of the thing patented, or a substantial part thereof claimed," &c., "judgment shall be rendered for the defendant with costs." Morse contends that the proviso to this last section, which says that patent shall not be held to be void if he believed he was the first inventor when he applied, modifies and controls this provision. I answer, he has stated no such case in his bill; and if he had, such an issue could not be tried. No one can prove what he believed at that moment. It would be impolitic to give a construction which would involve such an unprovable matter as his belief. The rights of others do not depend upon any such inscrutable facts. Under the old law, a patent could be decreed to be void, and it was then cancelled. The proviso is applicable to such a case. But there is nothing in this act for it to apply to, or operate upon. The Court is no where clothed with power to declare a patent void for any defects appearing upon a trial for its infringement. Its authority is confined to giving the defendant a judgment with costs when he proves certain defenses. It has no power to adjudicate the patent void, and that it be cancelled, or to do any thing else with it. This proviso was inserted in the Senate, after the bill passed the House, as the records show, and has nothing whatever to act upon, and serves no purpose but to confuse this most loose and imperfect law to be found in our statute-book, and one which, for the benefit of inventors and the good of the public, requires much amendment. I have high authority for my conclusions on the present point. Chancellor Kent, in his 2d volume of Commentaries, p. 568, (edition of 1848,) says: "If the patentee be not the first or original inventor, in reference to all the world, he is not entitled to a patent, even though he had no knowledge of the previous use, or previous description of the invention in any printed publication, for the law presumes he may have known it." In 1841, Judge Story, in Reed *vs.* Cutter, 1 Story R. 590, held, "Under the patent laws of the United States, the applicant for a patent must be the *first* as well as the original inventor." "It is sufficient that he is the first inventor to entitle him to a patent; and no subsequent inventor has a right to deprive him of the right to use his own invention. The language of the patent acts of 1836 and 1837, fully establishes this construction; and indeed this has been the habitual, if not invariable, interpretation of all our patent acts since the origin of our Government. The language of the act of 1836, 'not known or used by others, before his or their discovery thereof,' has never been supposed to vary this construction." "In Pennock *vs.* Dialogue, (2 Pet. 1,) the Supreme Court expressly held, that the section of the patent act then in force (and on this point the law has undergone no alteration) gives the right to the true and first inventor, and to him only; if known and used by others before his discovery, he is not the first, though he may be a true inventor, and that is the case to which the clause looks." Judge Woodbury, in 1846, in Allen *vs.* Blunt, 2 W. and M. 121, referring to the superior right of the first inventor, said: "There is no doubt it must prevail, as a matter of historical and chronological truth. It is in truth the first invention." If the law is as insisted by Morse, his oath to obtain his patents should have been, that he was the first inventor in this country, and his patent should have so stated the fact; but both oath and patent, in express words, conform to my construction. His bill should have conformed to his views. But his application, his oath, and his patent, the forms of the Patent Office, the words of the statute, and the adjudications as far as reported, are with me; and good common sense and sound policy are in harmony with them all. If Morse's position is correct, then if Steinheil should come here and use his invention, which was perfected and in actual use before Morse perfected his, he would be subject to a suit for using what his genius and learning produced and gave to the world. Wheatstone would be liable for using his combined circuits in this country, though actually patented by him in 1840, six years before Morse's application to patent them. Such a consequence shows the absurdity of Morse's pretensions. His patent is clearly bad.

In his fifth claim, Morse patents a system of signs. These are not covered by any provision in the patent law. If Morse can monopolize this kind of writing, he must resort to the copyright law. His pretensions that no one can combine dots, lines and spaces without violating his rights, cannot be sustained until he proves that he invented them. But we all know they are old, and it is so proved in this cause. He cannot combine these simple and common elements so as to prevent the rest of mankind making other combinations to meet their wants. This attempt at monopoly is only equalled by his claiming the exclusive use of sounds, without even showing how they can be used at all. For both these causes, this patent is invalid.

Eighth. Morse's reïssued patents, dated June 13, 1848, are invalid, because he has not shown that the surrendered patents were inoperative, or invalid, for defective specification or otherwise, so as to confer jurisdiction on the Commissioner to make such reïssues. The surrendered patents being set out, disprove any such jurisdiction.

Morse has not alleged in his bill, that his patents were "invalid" or "inoperative," nor that the error, if any, was occasioned by "inadvertency," or "mistake," or "accident," or "without fraud or deceptive intent on his part." Nor is it so stated in the reïssued patents. The patents say the surrender was "on account of defective specification." This does not show jurisdiction, while the surrendered patents, which are in the case, cannot be shown in the language of the act to "be inoperative or invalid, by reason of a defective or insufficient description or specification." The jurisdiction does not attach until the old patent is shown to be inoperative or invalid. The jurisdiction being statutory must be clearly made out. It is not pretended by Morse that his old patents were invalid, or inoperative. But he alleges, that the action of the Commissioner is conclusive. Where mere discretion is involved, this is so; but never in cases where the jurisdiction is in controversy. For certain purposes it has been held, that the reïssued patent is unreviewable. This has never been held upon the question of jurisdiction. But in this case the recitals in the patents fall short of making out the semblance of jurisdiction. There is no more reason why the action of the Commissioner cannot be reviewed in this, than in case of the original issue. Certainly this power is equally liable to abuse as the other. I have never known a surrender and reïssue where it was not apparent that the real object was something besides curing defects in the specification. As far as I know, the patentee has uniformly enlarged the boundary of his claims, although there is no stronger reason for it than for permitting the purchaser of a farm to enlarge the limits of his grant.

But I have the authority of this Court itself to sustain me. In Shaw *vs.*

Cooper, 7 Pet. 290, this Court said: "The same evidence which would defeat his application for a patent would at any subsequent period be fatal to his right. The evidence he exhibits to the Department of State is not only *ex-parte*, but interested; and the questions of fact are left open to be controverted, by every one who shall think proper to contest the right under the patent." In Grand *vs.* Raymond, 6 Pet. 218, it said, "If he shows that the patentee has failed in any of these pre-requisites, on which the authority to issue a patent is made to depend, his defence is complete." This case arose on a reïssued patent, and covers ours. The Court add, "The counsel for the plaintiff in error have shown very clearly, that the question of inadvertence or mistake is a judicial question, which cannot be decided by the Secretary of State. Neither can he decide those judicial questions on which the validity of the first patent depends. Yet he issues them without inquiring into them. Why may he not in like manner issue the second patent also?"

The correct performance of all those preliminaries on which the validity of the original patent depends, is always examinable in the court in which a suit for its violation shall be brought. Why may not those points on which the validity of the amended patent depends, be examined before the same tribunal? The rightfulness of issuing the new patent is declared (in the charge) to depend on the fact that the "defect in the specification arose from inadvertence or mistake, and without any fraud or misconduct on the part of the patentee." The jury of course were to inquire into the fact. The new patent was sustained because it stood this test. This case is like the one before us. The reasons assigned will equally apply, and will warrant this Court in reviewing the whole action of the Commissioner. On the question of jurisdiction, this Court has often expressed its opinion. In Walker *vs.* Turner, 9 Wheaton 541, this Court held that where the "jurisdiction is strictly special and limited," it was essential that "the record show that he had acted upon a case the law had submitted to him." In Wilson *vs.* Rosseau, 4 How. 646, this Court, speaking of the decision of the Commissioner in the case of an extension, said, "It is not conclusive upon the question of jurisdiction to act in a particular case." There are numerous cases to this point on the brief. I will only cite Lessee of Hickey *vs.* Stewart, 4 How. 750, where a former decision is approved and this rule established, "that the jurisdiction of any court exercising authority over a subject, may be inquired into, in every other court, when the proceedings are relied on, and brought before the latter by the party claiming the benefit of such proceeding."

These cases settle the question, that we may examine into the Commissioner's jurisdiction. Morse has failed to show the facts that conferred it, or any proof that he possessed it; but on the contrary, the only evidence upon the point in the case shows conclusively, that he had none. Not showing jurisdiction for these reïssues, these patents are invalid.

Ninth. Morse's patent of 1840, as secondly reïssued, is invalid, because the Commissioner had no authority to accept a second surrender and make a second reïssue.

These reïssues are the fountain-head of litigation, as well as the source of innumerable frauds. The presumption is, that the first reïssue cured all defects, if any really existed. There is no authority for a second reïssue, in the statute, which alone confers jurisdiction. Sound policy would prevent all reïssues, but as the statue provides for one, it should be confined to that. It is a well-settled rule of law, that when a special statue power has been exercised, it cannot form the basis of a second exertion of it. This was held in Demming *vs.* Smith, 3 John Chy. R. 332. Delafield *vs.* State of Illinois, 8 Paige, 527, and 26 Wen. 192, proceeded upon the same ground. The 13th Petersdorf, 644, lays down the same rule. Lucas *vs.* Attorney General,

11 Gill and John 490, and in Phalen *vs.* State of Maryland, 12 Gill and John 18, this doctrine was expressly held in Maryland. The changes and expansions that these patents present, must awaken the attention of the Court, as well as the public, to the impolicy, as well as the illegality of these repeated reïssues. If I am right on this point, the patent of 1840, as secondly reissued, is invalid.

Tenth. Morse's patent of 1840, as secondly reïssued, is invalid, because it is broader than the invention originally patented.

The fact that it is broader, is scarcely denied. That it is so, is proved, beyond controversy, by the record. In his reïssued patent, the first claim is for the exclusive use of the power of magnetism, which was not in the original. It was not then patented, nor did he invent it. It was then a claim limited to the application and arrangement of electro-magnets, which was a combination. In his new patent the second claim includes the employment of machinery, called a register, which is made to embrace a "fly" and "spiral spring." In the original, it was for a combination of machinery described, which did not include them. In the new patent, the third claim is for a combination, which constitutes the whole telegraphic apparatus. This includes the "fly" and "spiral spring," and breaking and closing the circuit by metallic contact, by applying the hand. These things were not included in the original. In the old patent, he claimed the "combination of electro-magnets in one or more circuits of conductors with armatures of magnets." In the new, he claims the combination of two or more galvanic circuits, with independent batteries, &c. This is an enlargement. The fifth claim in the new patent is for the invention of a system of signs, consisting of dots and lines. There is no such claim in the original. In the sixth, a system of signs, combined with machinery generally, is claimed; while in the original, it was for their combination with the machinery then described, which was useless. This is an enlargement, because it extends to all machinery, whether described or invented by Morse or not. It was doubtless intended by Morse to cover the machinery of the Columbian and House's Printing Telegraph. It is a palpable enlargement, and includes what Morse did not invent. The seventh includes in the combination the signal lever, or its equivalent, which is not in the original. The eighth claim for the exclusive use of the motive power of the electric or galvanic current, which he calls electro-magnetism, is wholly an addition, and was not invented by Morse. These are only a portion of the enlargements. Morse claims that they were a part of his original invention, and that they were lawfully included, if not in the original at all. We deny that they were a part of his original invention, and there is no proof whatever that they were so. The Commissioner had no such question before him, and no means are provided for its trial if it had been. Having failed to prove it, either before him or in this cause, by a scintilla of evidence, he cannot derive any advantage from it. But we deny that he has a right to enlarge his patent. The thirteenth section of the act of 1836 provides for the "adding the description and specification of any new improvement of the original invention," which shall have been subsequently invented. Most of these enlargements were, as seems palpable, additions subsequently made, and come within the words and spirit of this provision. The new patent, by statute, must be "for the same invention," which was so defectively described as to make the patent invalid, or inoperative. This is the only invention which the section had mentioned. The object is stated to be, to make the defective description a good one, so that it will actually confer rights; but it makes no allusion to the idea that new matter can be introduced, or new claims inserted. The section has no such object. The words "same invention" clearly mean the same that had been patented, though not properly described. If it had been intended to allow new things to be described and claimed, it would have been stated. To give it such a construction now, we must interpolate the words "originally made," where the word "patented" is most clearly implied. When an inventor assigns a patent for a machine, for the residue of his term, it would be unjust, nay unlawful, for the assignee to surrender the old, and take a new patent for things not included in it. The inventor may have intended to reserve the things not described, or claimed for his own use in another patent or otherwise. By assigning the old patent he confers no right to any thing not claimed and attempted to be described. Suppose an inventor assigns his application, and the patent is issued to his grantee. He conveys only what he claims and specifies. If he invented more, it remains his. What right has the grantee to what is not mentioned in the contract of sale? He did not purchase it. His contract or assignment should be interpreted, like all others, by the intention of parties to it, as expressed in the writing. No greater error can be committed than to apply rules in relation to patents, and things connected with them, which are not applicable to other and common transactions. Without giving a forced and unnatural construction to the words "same invention," this Court cannot hold that new claims and descriptions can be included in a reïssued patent. If properly construed, then these new claims and newly described things have been wrongfully inserted, and the patent is bad.

Eleventh. Morse's patent of 1846 is void, because material parts of it were in public use, and had been patented and described in a printed publication, before his application.

The first claim in it is for the invention of connecting circuits, so that one brings another into use. This was invented by Professor Henry, and used by him, as he swears, years before Morse claims to have made the invention. He also swears, that he saw it in use by Wheatstone, in April, 1837. It was patented by Wheatstone in 1837, and appears in Davy's, in England, in 1838, and is also described in the London *Mechanic's Magazine* in 1838, and by Vail (one of the plaintiffs) in his book published in 1845. Morse himself had patented it in 1848, in the second reïssue of his patent of 1840. It is there in these words: "I claim as my invention the combination of two or more galvanic or electric circuits, with independent batteries, substantially by the means herein described, for the purpose of obviating the diminished force of electro-magnetism in long circuits, and enabling me to command sufficient power to put in motion registering or recording machinery at a distance."

In the patent of 1846 it is, "the employment in a main telegraphic circuit, of a device or contrivance, called the receiving magnet, in combination with a short local independent circuit, or circuits, each having a register and register magnet, or other magnetic contrivances, for registering and sustaining such a relation to the register magnet or other magnetic contrivances for registering, and to the length of the circuit of telegraph line, as will enable me to obtain, with the aid of the main galvanic battery and circuit, and the intervention of a local battery and circuit, such motive power for registering as could not be obtained otherwise, without the use of a much larger galvanic battery, if at all." This long and foggy sentence covers a contrivance for connecting circuits and obtaining more power, which is the same thing claimed in the patent of 1840. By referring to the two specifications, they describe the same things, as far as described at all. Thorough examination will show that they use the same means and have the same description, and are employed for the same purpose. The registering and receiving magnets are simply Henry's electro-magnet. Although it is pretended that the former has stops and a spiral spring for adjustment, still these things are not described by Morse, nor do they form a part of the receiving magnet, nor do they constitute a patentable invention. But if they were described, they actually appear in figure two, example eleven, of the second reïssue of his patent of 1840, now in the hands of the Court, and therefore could not have been covered by another patent. If the old patent does not include his first claim in the new, then it is perfectly certain that he did not invent it until many years after Henry did, and after Wheatstone and Davy described it in their patents, and used it publicly in England, and until after it had been described in a printed publication in London. It has been alleged that the contrivance in the patent of 1849, as secondly reïssued, would not transmit the galvanic current both ways, by the use of a single wire. It would not, nor can the present contrivances do so. At first, Morse did not know the ground could be used as a part of the circuit, and hence a return-wire was needed and employed by him. But when he adopted this part of Steinheil's invention, the ground took the place of the second wire. But even now, the current cannot be transmitted both ways on the same wire without changing the poles of the batteries. The current always goes in one direction. If Baltimore wishes to send to Washington, the current leaves one pole of the battery, proceeds to Washington, and comes back to the other pole of the battery—going in one direction through the wire, and in the other in the ground. If Washington sends to Baltimore, the current leaves the like pole of the Washington battery, proceeds to Baltimore, and comes back to the other battery at the place of starting. It is the same in communicating on to Philadelphia and New-York. The pretence that there has been a change in this respect is all delusive. The only alteration by Morse, was the adoption of Steinheil's discovery of using the earth as a part of the circuit, without the grace of acknowledgment. But even this he does not describe and claim. This patent is invalid because of this long prior discovery, prior published description, and prior patenting, and prior use of the thing claimed.

Twelfth. Morse's reïssued patent of 1846 is invalid, because it is broader than the original.

This is the same question as the tenth proposition, being applied however to Morse's last patent. I shall, therefore, merely show the fact that it is broader. Morse, in his first claim in this patent, claims the employment of a receiving magnet, or its equivalent, in *combination* with a short, local, independent circuit, having a register magnet, for the purpose of obtaining more power. There is no such claim in the original. He there claimed the invention of the receiving magnet or one of similar character, sustaining certain relations to enable him to obtain power, without mentioning a short, local, independent circuit, or the present combination. He now claims two short local circuits in a new combination, which is a material enlargement of his pretensions. His third claim is enlarged by claiming indentations upon paper, and other fabrics, instead of using coloring matter. His historical recital, though not an enlargement of his claim, is an attempt at self-glorification, at the expense of prior inventors. If I am right, in relation to these enlargements, this patent is a legal nullity.

Thirteenth. The power to surrender and reïssue on account of *defective* specifications did not authorize Morse to make new specifications; and, having made *new* ones, his patents are bad.

This case illustrates the danger of allowing a patentee, under pretence of correcting and amending a defective specification, to make a new one. There was really no such defect as stated, and none can or will be pointed out in these old patents. The language has been changed so that the thing first patented cannot be followed, as the court will learn by trying to do so in the new ones. Instead of correcting a void thing, so as to make it valid, as is pretended in the statement in these patents, new inventions and enlarged claims are ingeniously introduced. The surrender was not made, nor the reïssue claimed or granted under a pretence of describing omitted things, or of enlarging the claims so as to include the whole invention, but to amend a defective specification. Under this pretense, in fraud of the law, a new specification is written out, and things not invented by him are introduced, and claims made to cover known inventions of others. Neither then nor now can Morse name one thing that was not sufficiently described before. The act speaks of the new patent being issued in "accordance with the patentee's corrected description and specification." In these cases, instead of a corrected, we have a new description and new claims. We ask this court to find, if it can, wherein these patents were inoperative or invalid by reason of a defective specification; and then to find wherein a correction is made which was necessary to secure their validity. Unless the words new and corrected are synonyms, this cannot be done. This is a fit occasion for this court to check this growing evil, by expressing its disapprobation of a pernicious practice, which in this case carries on its face the impress of fraud. For this cause, these patents should be held bad.

Fourteenth. The first and eighth claims of the patent of 1840, as secondly reïssued, are for a principle or effect, and not for a machine, manufacture, or composition of matter, or an improvement upon either, and it is therefore invalid.

This is the most important question raised in this cause. Its decision will determine whether our patent laws really promote the progress of the useful arts, or hold them stationary for periods of fourteen years; whether the principles of nature, not invented by man, can be monopolized by one to the exclusion of all others; and whether a patent can confer upon a patentee the exclusive right to use what he did not invent, and to make an inventor respond in damages for using his own invention. If this can be done under our present laws, they should be promptly changed, or expunged from the statute-book. Inventors cannot prosper under such a construction. They must be allowed the free use of what God has created, to mould and fashion it for man's use, or their energies will cramp and wither, and their faculties rust, without benefiting mankind, or aiding, as they have done, in the great work of the world's improvement. What the honest inventor desires, and is essential to his success, is protection of what he invents, and simple and effectual remedies for invasions, with prompt trials. The laws, in my opinion, should be so amended as to confine contests to questions of infringement and damages only. With our present loose laws and monstrosity trials, and unnatural and strange claims, under the law, if this claim to monopolize the principles of nature succeeds, the poor inventor can expect but little from his exertions to benefit himself or mankind. His inventions, and his chance of profits and hopes of honor, may be swallowed up by the prior monopolizer of nature's principles.

In his first claim he patents the exclusive use of the motive power of magnetism, and in the eighth that of electro-magnetism. I shall treat them as one, though they are essentially different. This was not claimed in his first patent. The dangers of competition were increasing. The fraud-path of a reïssue, with a skilfully-formed new specification, is resorted to. Knowing the kind wishes of the American people were with him, he was willing to run the hazard of patenting the exclusive use of the galvanic power through the electro-magnet. He now asks this court to sustain him in this fraud upon the law, and this bold intrusion upon the common property of man. He doubtless had heard that some English cases had held, that he who discovers a new principle, and has invented a mode of applying it, is entitled to its monopoly. This rule would serve his purposes, and the changes in his patent were made, and O'Rielly forthwith sued for using what was lawfully used the day before the reïssue. But he mistook the English rule. The English courts have never held, as I understand the decisions, that a power provided by the Ruler of the universe can be patented to or monopolized by any man. They have held that when an inventor has combined and arranged material things, whether powers, machinery, or other things, so as to produce something new and useful, the principle of action, or by which the effect is produced, is an invention and patentable; but not the simple elements or powers used. The English cases do not refer to a principle of nature, but to the principle of the artificial being or thing invented and patented—its mechanical, chemical, or other principles. Our adversaries do not cite one case which goes to the extent of patenting a principle of nature. Judge Kane's opinion upon Morse's patents has been cited. He held that Morse's patent of 1840 was for an art, not for a principle of nature, and that Morse had under it the exclusive right to such art for fourteen years. He held Bain's patent was for the same art, and therefore void. The case before him was one on Morse's patents of 1840, 1846, and 1849. It was claimed by Morse that Bain's was identical with his of 1849. Being identical with it, if Bain's was void, as covering the art patented by Morse in 1849, then Morse's must have been void also, because Morse could not re-patent his pretended art more than Bain. But Judge Kane held otherwise. He held Morse's patents of 1846 and 1849 to be valid, though they were for the same art he declared to be covered by the patent of 1840, and held Bain's to be void because Morse's last-mentioned covered the whole ground. This decision is a perfect *felo de se.* The identical principle which he applied to Bain's patent to kill it, if applied to Morse's two latest patents, would kill them also.

The case of *Battin* vs. *Taggart* was decided the same term at which that of *French* vs. *Rogers* was argued. If Judge Kane had applied in the latter case the same rule which I have before quoted from his decision in the former, Morse's reïssued patents must have been declared void.

He was equally unfortunate in his investigation of the facts. Morse's patent of 1846 does not show that the line can be worked both ways, nor that he patented and described an adjustable receiving magnet, nor does the record sustain the allegation that Steinheil's and Davy's inventions were mere semaphores, speaking to the eye. Both were recording telegraphs. Other errors could be enumerated. Having these things in mind, I cannot look upon the case as one of authority. Judge Kane went so far beyond the English cases as to stand alone. Our statute and the English are not identical. The English courts interpret their statutes, and our courts our own. But if our courts were to follow the English, they could not save this patent, because it is the principles of the machine, or the invention, and not those of nature, that they hold to be patentable.

But if we admit that he who discovers a principle, and shows a mode of using it, can have a valid patent to cover the whole principle, it does not help Morse in this case. In the first and eighth claims he covers the use of the motive power of electro-magnetism; but he did not discover that power at all. It was discovered and used before he attempted to contrive a telegraph. This is proved and also admitted by his vain recitals in his patents. He did not first discover the power, nor is there a pretence of it in the case. The proof is all the other way. He merely applied it, by attaching his recording apparatus. Hence this patent cannot be sustained on the ground that he first discovered the motive power of electro-magnetism. If there is nothing else to sustain it, the patent must fall. This claim is broader than his invention, and therefore bad.

It is next alleged that a recording telegraph is an art, and that Morse invented that art, and that he is entitled to a monopoly of it. This position is an abandonment of the other. But, unfortunately for those who urge it, this is not the thing patented. The first and eighth claims simply cover the use of a specified motive power. If these fail, the patent is bad. They vitiate it entirely. The other claims nowhere refer to this pretended art, and it is not set out in the specification. It does not find a resting-place in the patent at all. It only appears in my learned adversary's brief. The other side cannot point it out, and your Honors cannot find it in the patents. They will find only claims to dots, lines, types, combinations and machinery, &c., but no reference to the art, which, if it exists at all, must have been invented since the last reïssue. The next reïssue may include it, if the old expanding and grasping practice prevails.

Morse has an undisputed right to claim certain machinery, but he has embraced what he did not invent, which destroys the whole patent, as I have shown. But if we admit that Morse has claimed an art, has he invented one? Certainly not. "Art is the disposition or modification of things by human skill to answer the purpose intended." "Also a system of rules serving to facilitate the performance of certain actions." Has Morse, in this sense, invented an art; and if so, what art? The art of creating the galvanic current—the art of sending it from one pole of a battery, through a circuit, to the other—the art of making an electro-magnet—the art of attracting a needle or iron bar, or armature, by the electro-magnet, so as to create motion —that is, the motive power of electro-magnetism to indicate telegraphic signs—were all known and used before Morse's invention. The art of telegraphing includes things which he did not invent. All that he invented was indenting or marking these signs upon paper. This he did, having the aid of all the previous arts which I have enumerated. And in respect to recording, he was preceded by Steinheil. He cannot rightfully claim the art of telegraphing, because he only invented the manner of manifesting the messages, and not that of sending them. The whole means used for transmission were invented and used before his discovery. Hence the art of telegraphing is not his invention, and is not patented. The art of receiving the pulsations of the current in the electro-magnet was known and used before his invention. The attraction of the iron armature to the electro-magnet was not his invention. Henry and others had done this. The former attracted over three thousand pounds weight with an electro-magnet. The art of moving clock-work by weights was not new. Waiving Steinheil's older invention of marking by electro-magnetism, Morse did not wholly invent a mode of marking on paper by the motion of the armature, moved by the attractive power of the electro-magnet, because others had invented most of the elements and combinations used. The mode of combining circuits was invented by Henry and by Wheatstone before him. His only claim to the invention of the art of telegraphing is limited to a mode of marking, in which he was preceded by the learned Bavarian. If so shaping and suspending the armature as to cause it to rub against or strike upon moving paper constitutes on art, then Morse invented one; but not the art of telegraphing, because that includes more. If Morse has invented an art, it is not the one so vauntingly claimed, which includes numerous arts, or inventions, besides that of recording, which were perfected to his hands. The art of printing is referred to. Did any one man invent it? It is literally composed of more than a hundred inventions. Neither the paper-maker, ink-maker, type-maker, ball or roller-maker, nor press-maker, with the numerous inventions applicable to each, invented the art of printing. Even Hoe, with his wonderful press, cannot make claim to the art of printing. It is the aggregated labor of numerous inventors. So of the art of navigation, the art of making iron and steel, cutlery, watches, &c. It is equally true of telegraphing.

But if Morse had invented the art of telegraphing, his claims should be confined to his mode of doing it, and cannot extend to all modes, because he did not invent all modes. By law, he is bound to disclose all modes known to him, and a wrongful concealment would vitiate the patent. He gives one mode, and the presumption is, that he knew no other. His invention, it is well settled, can be no broader than his discovery, and hence can only cover his own mode. The English cases cited cannot apply, because there is no new principle involved. Electro-magnets had attracted armatures before—

weights had before moved clock-work to produce steady motion. Pens and pencils had previously marked, and dots, lines and spaces had before been used to signify ideas. Galvanic batteries had generated currents, and wires had conducted them. Circuits had been broken and closed by metallic contact. There is no reasonable pretence that any new power or principle was involved in what Morse invented. In truth, he has neither discovered a new principle in science or mechanics, nor invented the art of telegraphing, and was not the first inventor of a recording apparatus.

The American authorities repudiate the idea that any thing can be patented but the means invented to produce a result. In Wyeth *vs.* Stone, 1 Story R. 285, Story J. says, in relation to an invention for cutting ice, "The invention of this art, as well as the particular method of the application of the principle, are claimed by Wyeth. It is plain, that here, the patentee claims the exclusive art of cutting ice by means of any power other than human power. Such a claim is utterly unmaintainable in point of law." In Whittemore *vs.* Cutter, 1 Gal. 478, he said, "A patent in no case can be for an effect only, but for an effect produced in a given manner." In Stone *vs.* Sprague, 1 Story R. 270, he said, "We are of opinion that if the patent be construed to include all modes of communicating motion from the reel to the yarn-beam, and of the connection of the one with the other generally, it is utterly void, as being an attempt to maintain a patent for an abstract principle, or for all possible modes whatsoever of communication, although they may be invented by others and substantially differ from the mode described by the plaintiff in his specification. A man might just as well claim title to all possible or practicable modes of communicating motion from the steam-engine to a steamboat, although he had invented but one mode." The English cases of Burnton *vs.* Hawks, 4 Barn. and Ald. 541; Hill *vs.* Thompson, 8 Taunton 375; Boulton and Watt *vs.* Bell, 2 Hen. Bl. 487, and Losk *vs.* Hague, Webster, P. C. 207, are to the same effect. In the latter case, Heath J. says, "What he (the patentee) cannot specify, he has not invented." "This patent extends to all machinery that may be made upon this principle, so that he has taken a patent for more than he has specified." Buller J. says, "A patent must be for some new production from those elements, and not for the elements themselves."

When Morse's patents were before Judge Woodbury in Smith *vs.* Downing, this question was elaborately discussed and most fully considered. My brief gives the material portions of the opinion. The court then held that neither a principle, effect, nor an art could be patented. The Judge said, a patentee "can be protected only in the mode and to the extent described." Referring to English judges, he says, "Where any judge speaks of patenting an art, it is not the art in the abstract, without a specification of the manner in which it is to operate, as a manufacture or otherwise. But it is the art thus explained in the specification, and illustrated by a machine, or model, or drawings, when of a character to be. It is the art as represented, or exemplified, like the principle thus embodied, which, alone, the patent laws were designed to protect. In the English acts, the word 'art' is not used at all. And in ours, as well as in our Constitution, the word art means a useful art, or a manufacture which is beneficial, and which by the same law is required to be described with exactness in its mode of operation, and which, of course, for the reasons already laid down, can be protected only in the mode and to the extent thus described." The reasons given in this learned opinion were so strong that no appeal has been taken to this court from the decision. Judge Kane, in the Philadelphia case on these Morse patents, declared a contrary opinion, without doing himself or Judge Woodbury the justice to refer to the case at all, though cited by me. Judge Woodbury's argument was not then and cannot now be met and answered. It is true the learned judge construed Morse's patent to be simply for the mode which he described, and that it was not therefore for an art, or the monopoly of a principle, as is here contended. Holding it to be limited to the simple mode which I have stated, as all he invented, and for nothing else, he did not declare Morse's patent invalid, but held House's invention was no interference, because the mode was different. The patent of 1840, as now presented, has a claim for printing "signs *or* letters." In the one before him, it was "signs *of* letters." As this claim is an interpolation upon the original, and covers marking or printing letters, and there being not even the pretence of description for producing them, this patent must be invalid on that account. Had it appeared in this form before Judge Woodbury, he, doubtless, would have held the same way. The principal reason usually assigned for sustaining these patents, so full of legal faults, is that the patentee has benefited mankind by what he has done, and ought to be rewarded. If the evidence did not show that benefit to himself was the ruling passion, I might answer, that thousands benefit mankind by their inventions and discoveries, where no law confers on them a pecuniary reward. Did Franklin patent his discoveries, or Lieut. Maury his? Professor Henry, who invented the heart, the very soul of Morse's telegraph, has neither asked nor received a pecuniary reward, while Morse uses it as freely as he does his own, which would be utterly useless without it. My great objection to Morse's patents is, that he attempts to monopolize one of nature's elements long known and used. He seeks to withdraw it from common use, not because he first discovered it, but because he has used it in a particular way. He tells the inventive world to beware! this power is mine; I have secured it in my reïssued patents, and you use it at your peril. Suits at law ,suits in chancery, injunctions and ruinous litigations are threatened against those who dare now to use a power which others used long years ago. If the record shall be made by the action of galvanism upon paper saturated with certain salts, where no motive power is used at all, Morse claims that it violates his monopoly, and Judge Kane sustained him in the claim. In the present case, the use of the power of permanent magnetism is alleged to be an invasion of his patent, and the judge below sustains this extravagant proposition. Morse's pretensions are not limited in practice to the use of the motive power of electro-magnetism to mark; but if used for any purpose whatever, however remote from his manner—if a mark is made, even by one who invented the mode—the cry of infringement is heard, and the grapples of the law fastened upon him. Let a new invention appear, a surrender and reïssue enables Morse to include it. The inventions of Barnes and Zook, involved in this case, of Bain's in the Philadelphia case, and of House's in the one at Boston, were all sought to be brought within Morse's expansive and growing patents by the legerdemain of reïssues. House was prosecuted for using his own improvements, which Morse never invented. Rogers, assignee of Bain, was enjoined from using Bain's improvement, of which he was the first inventor. Numerous other suits are pending, which renders the decision in this case doubly important.

It is error to suppose that Morse has been personally involved in, or incurred heavy expenses for litigation. In this case he shares it with his co-plaintiffs. In all the other cases, so far from his being a party, or involved in expenses, others are plaintiffs, and he has uniformly appeared as their witness. Your Honors are now called upon to determine whether Morse can rightfully prosecute others for using what he did not invent—for using their own inventions or those of third persons. If Morse has the exclusive right to the motive power of electro-magnetism, as he claims, it follows that if I make an invention a hundred times superior to his, he can lawfully use my invention, and punish me if I use the fruits of my own genius and toil. He can recover for what he did not invent or even describe in his patents, and prevent American citizens using what they or others found out and he did not. It would hardly be safe for Henry to use his electro-magnet if a record resulted from it. In his first patent of 1846, Morse actually claimed it by name. Steinheil and Davy, if not Wheatstone, might be punished if they should here use their own prior inventions. The chemist and electrician must abandon their laboratories, if telegraphs occupy their thoughts, and the arts and sciences must come to a halt for fourteen years, or the proprietors of this undescribed and indescribable and boundless monopoly may open the batteries of the law upon them and demolish their hopes of fame and reward.

Most owners of property can describe their boundaries, but Morse's claim is too broad for description. It covers the productions of all mankind. It might well be called a patent to restrain invention and punish inventors and those patronizing them. To sustain it would be a most fatal blow to the inventive genius and energies of the country, and would defeat the policy of our Constitution and laws, and render them worse than useless.

www.ingramcontent.com/pod-product-compliance
Lightning Source LLC
LaVergne TN
LVHW011141110826
845150LV00008B/2445

* 9 7 8 1 4 1 8 1 9 2 9 4 5 *